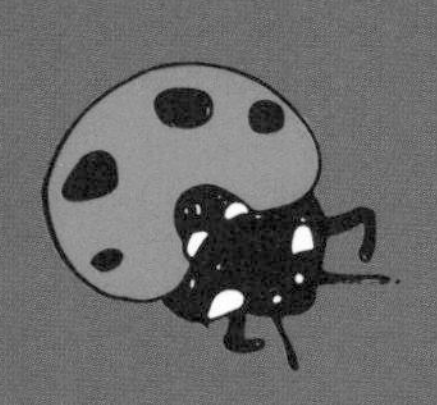

GW00646523

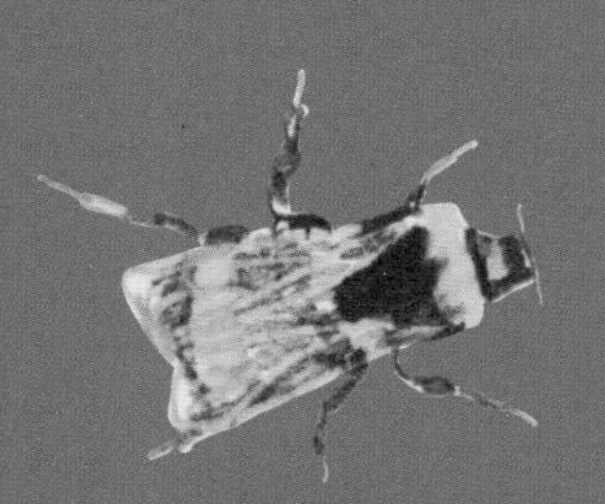

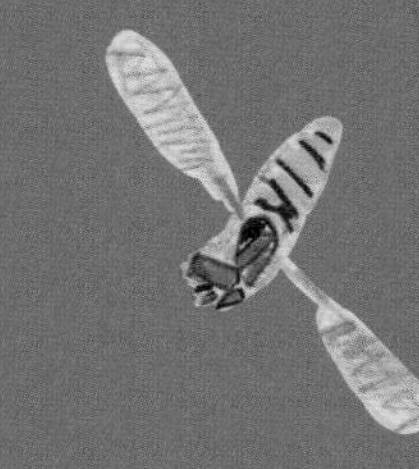

NATURAMA

Michael Fewer WITH Melissa Doran

GILL BOOKS

DEDICATED TO MY CONSULTANTS:

Finn, Diarmuid, Kerry, Dearbhla and Darina

CONTENTS

INTRODUCTION

In this book I will tell you about the animals, birds, insects and plants that you can see outdoors, if you look carefully, in each of the four seasons of the year: **SPRING**, **SUMMER**, **AUTUMN** and **WINTER**.

It is important to remember that we, and every living thing in nature, depend on one another, and on the air, water and soil of the earth for our health and happiness. We all need one another. For this reason, we should avoid harming the resources and creatures of our world in any way and, indeed, we should all do our best to take care of them.

SPRING

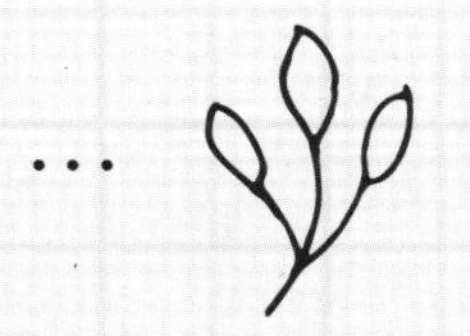

AT THE BEGINNING OF SPRING, TINY LITTLE GREEN SHOOTS BEGIN TO APPEAR ON PLANTS, AND THE BRANCHES OF TREES PUT OUT BUDS THAT FATTEN EVERY DAY AS THE BABY LEAVES GROW INSIDE THEM.

...

BIRDS ANNOUNCE THE end of winter with their songs, and you will hear the blackbird filling the air with sweet tunes in the evenings. Spring starts slowly. Ireland, which has been tilted away from the sun for the winter, begins to incline towards the sun, and the days start to get longer and warmer. Although there will be some cold and frosty days in spring, it gets warmer as the weeks go by, and you will notice that it is still quite bright at bedtime.

In ancient Ireland, people watched carefully for the return of longer days and warm sun, which would allow them to grow food crops again. No crops would grow in winter, so food had to be stored in autumn to last until spring. It was a worrying time, because there would be very little left by the time spring came, and there were no shops where you could buy food. Everyone must have been happy when the days began to get longer and warmer, and plants started to grow again. A big party was held to welcome St Brigid's day (1 February), the first day of spring. Today, spring officially begins in Ireland on 1 March and ends on 31 May.

Let's go outdoors and see what nature is doing in spring.

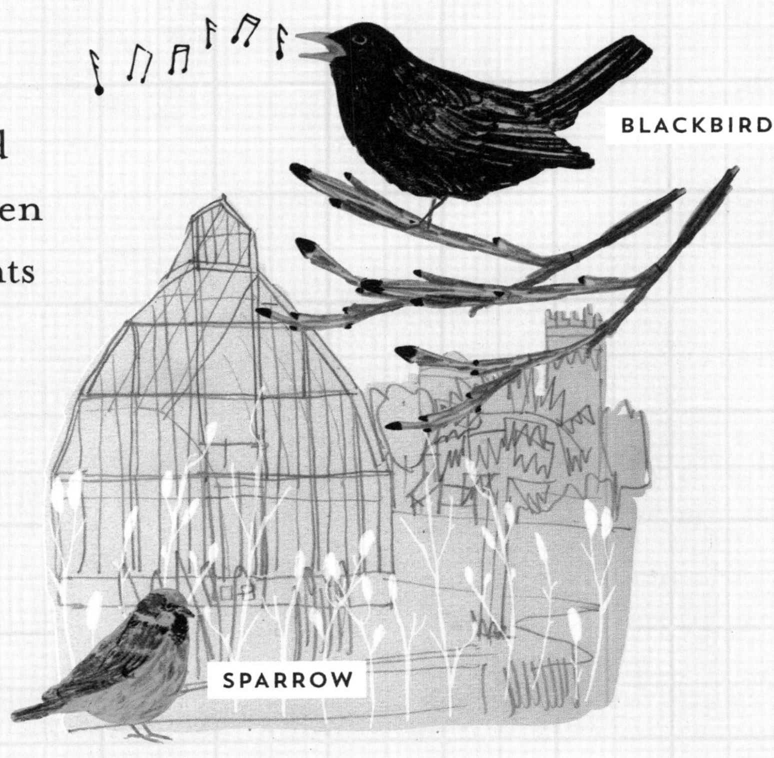

WHAT YOU CAN SEE IN

The Garden in Spring

AS THE WEATHER gets warmer, insects and other tiny creatures that have been hibernating or sleeping in warm, dry places over the winter begin to wake up and start to hunt for food. Many insects will have laid eggs in the autumn, and these start to hatch in spring and fill the garden with lots of new creatures. Have a look under stones and flowerpots to see what you can find.

The most common creature you will find in early spring is the **WORM**, which enjoys dark, damp places. Later, it will head for flowerbeds and lawns to burrow deep, away from the summer heat and dryness. Some children love worms, while others find them disgusting! Worms, however, are very important creatures. They clean up the soil by feeding on dead vegetable matter, and their burrows bring air into the soil, making it healthy and fertile, and good for growing grass, plants, trees and crops.

Worms might be small, but there are many millions of them under the ground. This is good, because without them not many plants would grow and we would have little to eat.

Under flowerpots you can find **WOODLICE**, creatures with 14 legs and grey armoured shells on their backs to protect them. Woodlouse in Irish is *míle críonna*, which means 'a thousand wrinkles'! Woodlice are very useful creatures for keeping the garden tidy because they like to eat decaying plants, leaves and rotting wood.

In late spring a creature that is easy to spot in the garden is the **LADYBIRD**, a tiny red/orange beetle with black spots. The bright colour of lady-birds is a warning to birds and frogs that might want to eat them that they have toxins in their blood, which taste awful. These little beauties are in fact predators, and they are useful in the garden because they attack, kill and eat pests like **GREENFLIES**. Greenflies are little green creatures with wings and fat green bodies, and they like to eat rosebuds, preventing roses from growing.

...

ONE OF THE earliest flowers to appear in spring is the **CROCUS**, a plant that was brought to Ireland many hundreds of years ago from Greece. There are many different types of crocus in a range of beautiful colours.

The **DAFFODIL** is another popular flower that is not native to Ireland, but was brought here centuries ago from

GOLDFINCHES

Africa and south-west Europe. When you see daffodils waving their yellow and golden heads in the breeze, it is a sure sign that spring has arrived. Children used to sing:

Daff-a-down-dill
Has now come to town,
In a yellow petticoat
And a green gown.

...

FROM THE BEGINNING of spring the birds will be busy about the garden, filling the air with their song. Male birds will trumpet and throw shapes in order to attract the females. Spring is the time to find a mate, build a nest and have young chicks, and, now that winter is over, lots of food is available to feed them. Birds spend their days looking for mates and searching for good places to build their nests – somewhere sheltered, dry and safe from predators such as cats. They build their nests from twigs, grass and straw, and glue them together with spiders' webs, mud and their own spit! Nests are usually lined with any soft materials they can find, such as moss or sheep's wool.

When the young birds leave their nests and start to learn to fly, often when they are only a few weeks old, they face a dangerous time. If they don't learn to fly very quickly, they could become an easy meal for a passing cat or for bigger birds such as the magpie and the sparrowhawk. The parent birds try to keep their chicks together in the garden and continue to feed them until they are able to look after themselves.

The **ROBIN**, with its tinkling song and its orange-red breast, is the easiest garden bird to identify. It is always flitting about or standing on a branch, watching you with tiny sparkling black eyes. If there is any digging or planting going on in the garden, the robin is sure to be there, always alert for a tasty worm or insect to be dug up. The female and the male robin have the same colouring and a red breast, but the female is usually a little smaller.

Robins build their nests in thick ivy or holes in walls, or even in flowerpots. The female lays five or six light grey eggs, speckled with red.

Other colourful little birds called **TITS** can be seen in gardens, the most common of which is the **BLUE TIT**. Blue tits nest in trees, holes in walls and in funny places like old letterboxes! They lay about nine or ten eggs. Often the chicks will have learned to fly just three weeks after they have hatched.

The champion songster of spring is the male **BLACKBIRD**, who finds a high place on a tree to sing his melodic song to attract a mate and to tell other male birds that this garden is his territory. He is black with a bright yellow beak, but the female has brown feathers, the same colour as the undergrowth where she builds her nest. This colour makes her almost invisible to predators such as cats when she is sitting on the nest hatching her eggs.

Have you seen any of these birds in your garden?

BLACKBIRDS

THE FOX USED to live only in the Irish countryside, but now is very common in towns and cities. The **FOX** is called *an sionnach* or *an madra rua* in Irish, meaning 'the red dog'. They feed on small mammals, insects and birds, and in towns they will also enjoy left-over food from bins. Because they usually leave their dens to hunt or get food at night, we don't often see them in daylight, but at the beginning of spring you may hear them in the quiet of the night. The female fox, called a vixen, makes a scary sound like a scream!

WHAT YOU CAN SEE IN

The Park in Spring

AT THE BEGINNING of spring the trees in the park are still bare, but by the middle of March the buds on their branches are swelling. Buds are little packages in which leaves and flowers are tightly folded until the time comes for them to open. Take one off a branch and open it up to see what you can find inside. The buds of the **HORSE CHESTNUT** tree are brown, sticky and large because chestnut leaves and flowers are really big.

Trees, like all plants, need sunlight to help them grow, and as the days get longer and the air and the soil become warmer, their buds begin to open and the leaves unfold and expand. As the leaves expand, the branches of trees are gradually hidden and the tree turns green all over.

The **ASH TREE** is one of the last of the trees to put out its leaves. Ash is a very strong wood that is used for making furniture and hurleys. Hurling is sometimes called 'the clash of the ash'. In ancient times the warrior Queen Maeve had a horse whip made from ash wood, and wherever she put it down, a magic ash tree grew! Three of the five legendary Great Trees of Ireland were ash trees: *Bile Uisnigh*, *Bile Tortan* and *Craobh Daithi*. *Bile Uisnigh* grew at a place called Uisneach in County Westmeath, at the ancient centre of Ireland. *Bile Tortan* grew at Ardbreccan in County Meath. *Craobh Daithi* grew in Farbill, also in County Westmeath, halfway between the other two places.

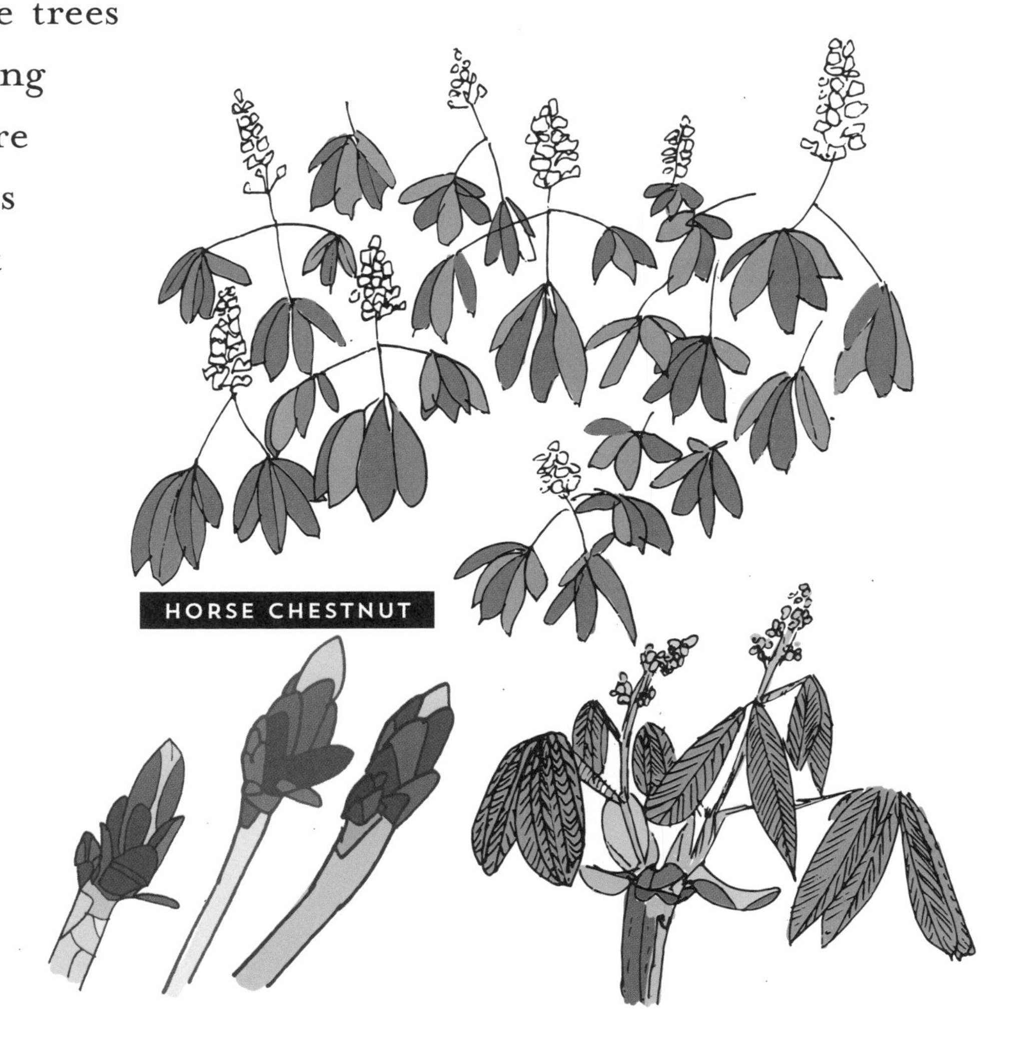

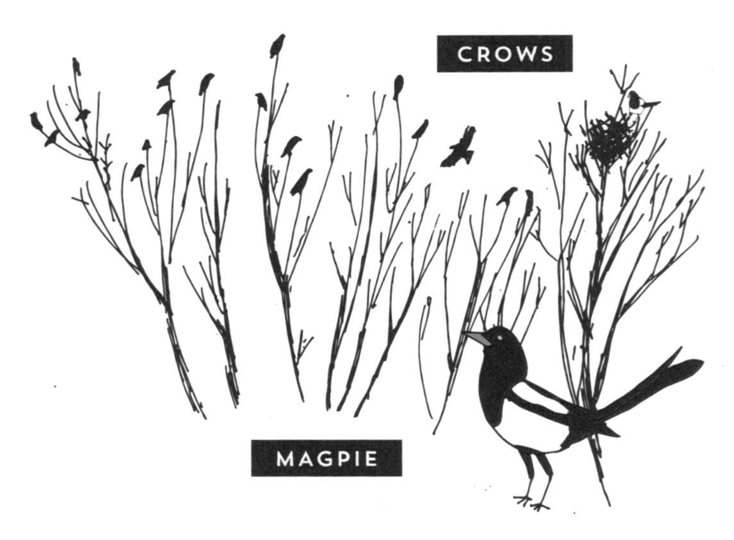

ROOKS, **JACKDAWS** and **MAGPIES**, all members of the crow family, build big nests in spring out of large twigs and sticks, high up in the branches of trees. You will often see them flying with large sticks in their beaks. In their nests they make a soft bed out of moss and feathers and wool for their young. They are untidy builders, and many of the sticks they gather fall to the ground.

Lots of parks have large ponds where you can see birds that like to live on or near water. In spring they are also busy building their nests.

The **WATER HEN** is a black bird about the size of a pigeon. It has a red forehead, a yellow beak and white feathers in its wings. The female builds her nest out of dry, dead rushes, usually among the reeds at the water's edge. She is quite shy, but if you go near the nest she will make a loud 'crek-crek' call to try to scare you off. As soon as her young chicks are hatched, they go swimming with her on the pond.

The **MALLARD** is Ireland's most common duck and can be seen on most ponds. The male has a bottle-green head, a yellow beak, a white collar and a grey and brown body. The female is speckled brown; when she is sitting on her eggs, it is hard to spot her. Male mallards have noisy arguments during the mating season in spring, and have loud splashing chases on ponds.

Mallards feed on lots of things, including insects, snails, frogs, seeds and water plants. They will flock to the water's edge if you offer them food. Don't feed them bread because it is not good for them and can harm the baby mallard's growth. If you want to feed them, it is better to offer them good, nutritional foods such as rice (cooked or uncooked), peas and wheat or barley grains. Try not to leave food scattered about the edge of the pond because it will attract rats and other pests that may be harmful to the birds.

If you want to feed the **SWANS** in the park, they like lettuce, spinach and chopped or shredded carrots.

THE EARLIEST FLOWER to appear in spring is the **CELANDINE**. Its flowers have shiny, yellow, spiky petals. Celandine grows in sheltered sunny patches in hedges and woods. The poet William Wordsworth wrote of it:

There's a flower that shall be mine;
'Tis the little celandine.

In the woods you can find another spring flower called the **WOOD ANEMONE**. This little plant must bloom before the leaves come on the trees because, after that, it will not get enough sunlight. Its spiky leaves are divided in three and its flower has white petals and yellow stamens. In spite of its delicate beauty, the wood anemone should not be picked because it is poisonous! Although it is always found in sheltered woodland, in Irish it is called *lus na gaoithe*, or 'the plant of the wind'!

WOOD SORREL is a similar plant, found also in shady places in woods towards the end of spring and the beginning of summer. Its white flower has five petals with tiny purple veins, and its leaves are divided in three, are round and look like large shamrock leaves. In ancient Ireland, people ate wood sorrel as a salad, and it was also used as a medicine.

WHAT YOU CAN SEE IN

The Countryside in Spring

THE INSECTS THAT are the easiest to see in the countryside in late springtime are **BUTTERFLIES**. Some of our Irish butterflies find a dry warm place, often in houses, to sleep through the winter. They come out again as soon as the spring sunshine warms the air. If you find them fluttering at a window, try to let them out without harming them. Other butterflies that appear in spring may have flown all the way across the sea from Britain to stay in Ireland for the spring and summer. They have an unusual life. They lay their eggs on the leaves of a plant, but when the eggs hatch it is not butterflies that emerge, but **CATERPILLARS**, and they immediately begin to eat the leaf on which they hatched. As soon as the caterpillar has grown to its full size, it turns into a creature called a pupa, which grows a hard outer skin. Inside this skin it changes again and grows wings, and then it leaves the hard outer skin behind and comes out as a butterfly.

Another insect that can be heard as well as seen in spring is the buzzing **BUMBLEBEE**, which has a round body covered with fur. There are a number of different kinds, but most are black or brown with yellow stripes. Bumblebees are very important for farmers who grow strawberries, tomatoes and apples, because they transfer pollen from one plant to another while they are feeding on the plant's nectar, and this allows the plants to produce fruit. Without bumblebees we would have very little fruit. Bumblebees have a sting, but they use it only if they are being attacked or threatened.

How many types of bumblebee can *you* find?

By April you can find masses of bright yellow **DANDELIONS** along the road verges. The dandelion used to be called 'The Sunflower of the Spring', and poor people in France used to eat its leaves in a salad. Each flower head contains up to 300 separate tiny flowers, and each little flower becomes a seed in late summer with a little parachute of silky hairs. In the old days children used to try to tell the time by the number of times they had to puff to make all the little seeds fly away. Each seed

will make a new plant, so you can see why there are so many dandelions.

Dandelions also grow in gardens, but because they are a weed, people don't like them. They do a lot of good, however: their little flowers are bursting with nectar, providing an excellent breakfast for hungry bumblebees when they come out of hibernation after the winter. The bumblebees thengo on to do their important job of pollinating fruit crops. Because too many wild flowers, such as dandelions, are being destroyed by gardeners, our bumblebees are dying out, and Irish farmers who grow strawberries and tomatoes have to buy bumblebees from Germany to keep their plants going!

...

IN THE COUNTRYSIDE in Ireland most roads and fields are bordered by hedgerows, which are banks of earth or stone walls covered with a rich mixture of trees, shrubs and plants. Hedgerows are important for nature because they provide good places for many plants to grow which would not survive in the fields, and they are comfortable homes for lots of animals, birds and insects. Hedgerows are full of life in spring. Birds build their nests in thick, brambly parts, and field mice and shrews have their homes in burrows down below. There is usually a rich mix of plants in a hedgerow, including the **WILD ROSE**, **HONEYSUCKLE**, **HAWTHORN**, **BLACKTHORN** and, of course, the thorny **BRAMBLE**, which gives us blackberries in autumn.

Some of the shrubs and trees of the hedgerow produce blossoms or catkins before their leaves grow. The **WILLOW** is a small tree which produces beautiful furry catkins in spring that provide early nectar for bees. In the old days thin willow branches were used to make baskets, and harps were made out of the wood of the willow tree. The most famous, Brian Boru's harp, can be seen in the Old Library of Trinity College Dublin.

The **PRIMROSE** has crinkly leaves and clusters of beautiful pale yellow flowers. It grows in hedges and grassy banks and its flowers are very fragrant. The primrose was linked in the old days with May Day, when it was gathered by children before dusk on May Eve, the day before 1 May. It was made into posies and hung over the door of the house to protect the family from fairies and goblins. It was also used to make a tea for people who couldn't sleep at night, and it was said that toothaches could be relieved by rubbing the sore tooth with the leaves of the primrose!

Shouldn't every garden have a little wildflower corner?

...

LATE IN THE month of May, in damp places in Irish bogs and mountains, you can find plants that 'eat' insects.

BUTTERWORT

Because these plants grow in places where there is not sufficient good food for them to take from the soil, they have learned to add insects to their diet. When an insect climbs onto the plant looking for food, it gets stuck in a sticky substance and the plant slowly closes up the leaf on which the insect is stuck. The insect is then turned into liquid on which the plant can feed. These plants are called insectivorous plants and we have two species in Ireland. The **BUTTERWORT** has narrow, pale green leaves and a beautiful purple flower. In ancient times people believed that the butterwort flower sprang up everywhere St Patrick placed his staff when he was walking across a bog!

The **SUNDEW** is another insectivorous plant. It looks like a tiny green and red cactus with glistening beads on its leaves. Long ago it was boiled in milk as a cure for whooping cough and asthma.

One of the most beautiful sights of late springtime is a woodland floor carpeted with **BLUEBELLS**. Called the 'Sapphire queen of the mid-May' by the poet John Keats, the bluebell has a deep blue/violet colour, and also has funny nicknames such as 'Crowtoes' and 'Cuckoo's boot'. In the old days the bluebell had many uses, including making a juice for curing sore throats and producing a glue used to stick feathers on an arrow!

BLUEBELLS

IN FEBRUARY **FROGS** gather, often in great numbers, at the nearest pond for their annual mating 'festival'. It is one of the most exciting things to see in spring, with hundreds of frogs crowded in the water with their heads just sticking out, singing, 'ribbit, ribbit, ribbit'. The female frogs produce lots of frogspawn, a colourless jelly with thousands of eggs that look like little black dots in it, and it floats in the water. The eggs will slowly develop into tadpoles, which resemble little black fish. Frogs lay thousands of eggs because they don't look after their young after they are born. Most of the little tadpoles don't survive because there are many predators such as fish, newts, water beetles and dragonflies in ponds, which like to feed on them. Only a very small number will survive to become adult frogs, growing back and then front legs before they climb out of the pond. When the mating festival is over, the frogs leave the pond and spend the rest of the year in dark, damp places. Frogs can live for about five or six years. They return to the same pond to attend the frog festival each spring!

Frogs have their own prey – usually snails, worms, insects and spiders – which they catch on their long sticky tongues. Their tongues can spring suddenly out of their mouths, giving their prey no time to get away. Frogs have no teeth, so they swallow their prey whole! During the winter they hibernate under rocks or stay buried in mud at the bottom of a pond.

In the old days frogs were regarded as creatures of the underworld and were thought to be friendly with witches. It was said that a sore eye could be cured by getting someone to lick a frog's eye and then lick the sore eye. Ugh!

...

ON THE HILLS and heathery moorland you can see one of Ireland's most common birds, the **MEADOW PIPIT**. It is small and brown and has a speckled breast. Most brown birds look much the same and are difficult to name, but the meadow pipit is easily recognised because in spring it flies 12 or 14 metres into the air singing 'cheep, cheep, cheep, cheep' all the while, and then comes down with its trembling wings spread like a parachute. It lays its five or six eggs on the ground in the shelter of a tuft of grass. Pipits sometimes protect their nests by pretending to have a broken wing, and flop along the ground leading the person away from it.

In the countryside you may also hear the harsh squawk of the male **PHEASANT** and, if you are lucky, see the large colourful bird with its long tail. They were brought to Ireland from the Far East many years ago by big landowners so that they would have interesting birds on their land to shoot!

One bird that you can hear in the countryside is the cuckoo, whose call is just like its name. The cuckoo is a strange and lazy bird that never makes its own nest, but instead lays its eggs in another bird's nest. The other bird doesn't seem to notice, and hatches the young cuckoo with its own chicks and even feeds it!

...

SPRINGTIME IS WHEN new young animals are born on the farm. Calves are born on dairy farms. Cows are well fed with new grass and can produce lots of milk for their new-born calves.

LAMBS are usually born indoors and then they are let out in the fields. They get strong quickly, drinking their mother's milk and eating the fresh spring grass. It is very funny to watch young lambs playing with one another, gambolling, running and chasing.

In the air over open farmland you may spot a **KESTREL**. The kestrel is a raptor, a bird that preys on and eats other birds, animals and insects. It is a beautiful bird with a speckled brown back, grey head and hooked beak. It is an expert at hovering, like a helicopter, in the same position, even in high winds, while it uses its remarkable eyesight to search the ground for its prey. It can even see a beetle on the ground from 50 metres up in the air!

SUMMER

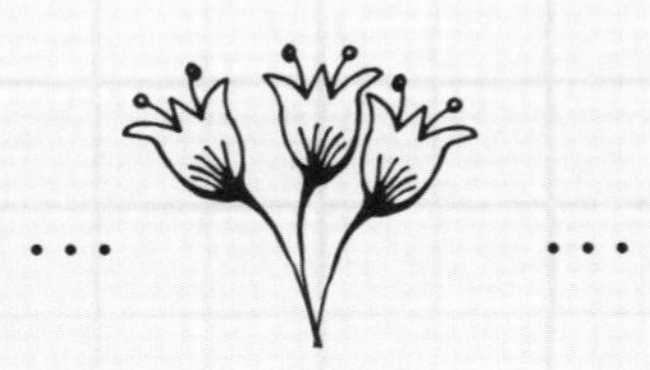

... ...

ALTHOUGH SUMMER BEGAN TRADITIONALLY ON 1 MAY, TODAY SUMMER OFFICIALLY BEGINS ON 1 JUNE AND ENDS ON 31 AUGUST. DURING THE THREE SUMMER MONTHS, THE NORTHERN PART OF EARTH, WHERE IRELAND IS SITUATED, IS TILTED TOWARDS THE SUN AND THAT IS WHY SUMMER IS THE WARMEST OF OUR SEASONS.

...

IN SUMMER, DAYS are long and nights are short. June is the month of the strongest and brightest sunlight, and the longest day of the year in Ireland is Midsummer's Day, 21 June. In ancient times the people used to celebrate Midsummer's Eve by lighting large bonfires on surrounding hillsides in honour of Baal, the sun god.

Summer is probably your favourite season because you have long holidays from school, and these months are the best time for playing sports, going to the beach, swimming in the sea and having picnics.

Nature is at its most active in summertime, and it's the time when the largest number of insects, animals and plants are busy in our gardens, parks and in the countryside. Summer warmth brings blossoms to plants and shrubs, and fruit trees and bushes produce flowers that in autumn will become delicious fruit, like apples and blackberries. Gardens and parks are decorated with colourful and fragrant flowers, and the air is full of butterflies, bees and hoverflies. It's a great time for looking at nature, so let's go out to the garden, the park, the seaside and the countryside and see what we can find there.

WHAT YOU CAN SEE IN

The Garden in Summer

THE GARDEN IN summertime is full of activity, colour and sound, with lots of insects and other creatures creeping and flying about.

The common **CENTIPEDE** is chestnut brown and has 30 legs, rather than the 100 that its name suggests. These little creatures are fierce predators. They attack smaller insects, killing or stunning them with their two front legs, which have sharp venomous claws, before eating them. The **MILLIPEDE** is a bit smaller, but it can have as many as 400 legs! It feeds on decaying leaves and other plant matter. Scientists who study millipedes are called diplopodologists.

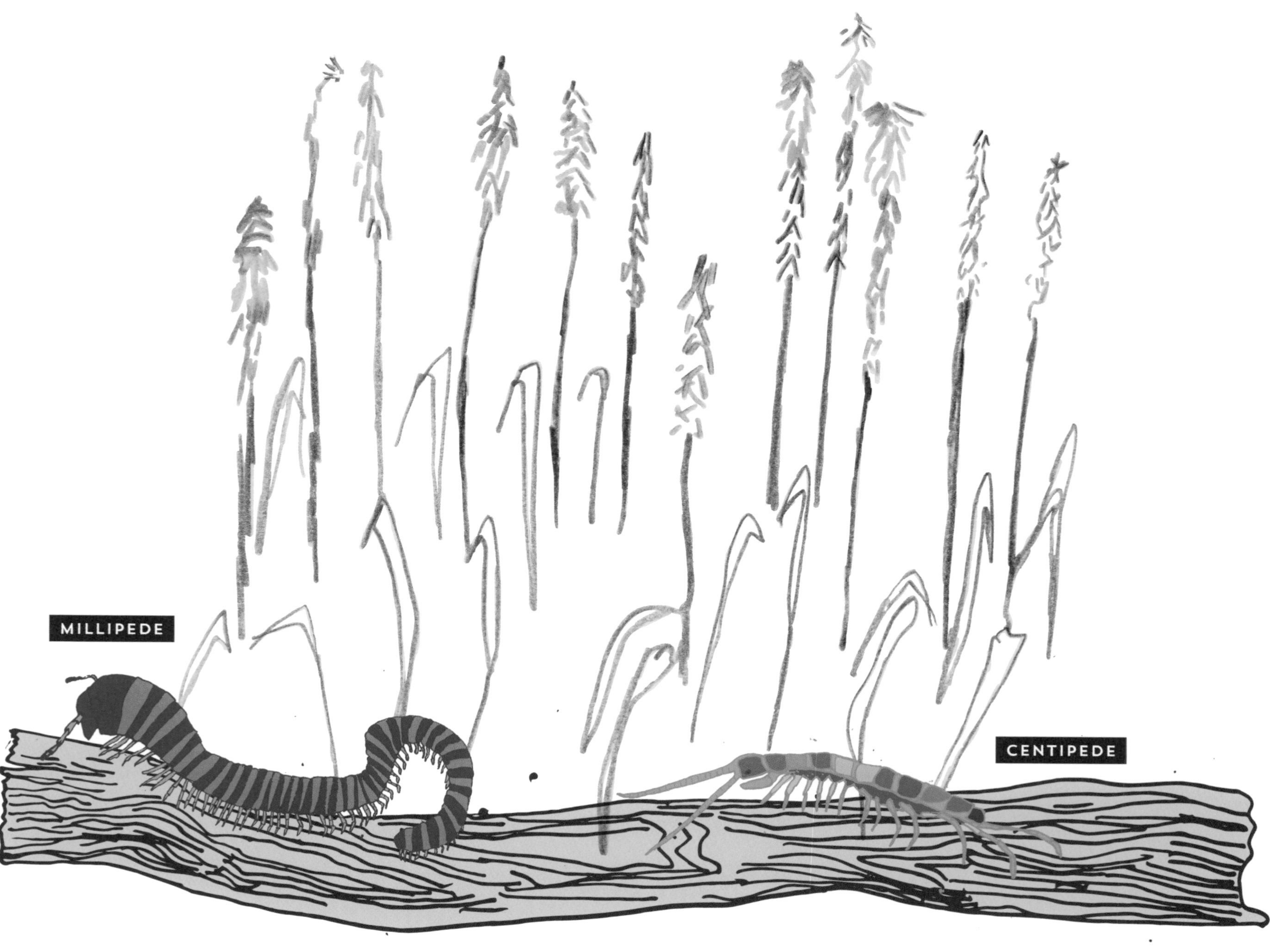

The **SPIDER** not only has eight legs but also eight eyes. They are plentiful in the garden in summer and they kill their prey by gripping them with their fangs and injecting poison. There are many different kinds of spiders. Some catch their prey using sticky webs, others creep up on their prey and jump on them, and there is even an indoor spider that spits a gummy web at small insects, sticking them to the ground. Hunting spiders come out at night and usually catch woodlice, earwigs and beetles. Irish spiders are harmless to humans.

The common garden spider spins a circular web like a net, made from silk strands, to catch little insects flying by. The silken web is almost invisible, but it is stronger than a strand of iron of the same thickness.

Insects are creatures with six legs, and there are many of them in every garden.

The **BEETLE** is one type of insect you will find in the garden. Beetles have a hard shell to protect their bodies, and most of them, such as the pretty ladybird, can fly. The most common garden beetle is the shiny black ground beetle. This beetle can be found under stones and logs, and runs fast to get away if it is disturbed.

The **HONEY BEE** has a brown furry body and feeds on nectar, a sweet liquid produced by flowers. Honey bees live in hives that are provided by people called beekeepers. In the hives the honey bees

HONEY BEE

convert the nectar they take from flowers into honey. The beekeeper takes the honey from the hive in mid-August and puts it into jars to be sold. As well as being deliciously sweet, honey is a very healthy food, rich in vitamins and minerals.

The **BUMBLEBEE** is larger and fatter than the honey bee, and usually has a black or brown body with yellow stripes. They can sting, but usually only do so when their hive is being attacked.

All bees are important pollinators, and crops and fruits could not grow without them. To make sure there will be new young plants after they die, most plants produce a fine powder called pollen in their flowers. But each plant has to mix its pollen with pollen from another flower to create the seeds that will produce new plants. They use bees and other insects to bring their pollen to nearby flowers, cleverly attracting them by also producing nectar. Bees and insects love nectar, and as they climb into the flower to get it, some pollen rubs off on them. When they visit another flower to drink its nectar, the pollen drops off, allowing that flower to produce seeds for new plants. Without pollinators, there would be no new plants after the old ones die, and we would have none of the fruits or vegetables that are an important part of our healthy eating.

Another insect that pollinates flowers while it drinks nectar is the **HOVERFLY**. It has coloured stripes on its body and pretends to be a wasp to scare off predators! You will know it is a hoverfly when you see it hovering in one place like a helicopter. Hoverflies are harmless and do not sting. They are among the 'good' insects in the garden because they feed on **GREENFLIES**, which cause damage to plants.

WASPS have yellow bodies with black dots and stripes. They build hives made from paper that they make themselves by chewing little bits of wood and mixing it with their saliva. The hives are beautiful and can be found in old roofs, attics, trees and sheds. Wasps' hives often look like pretty paper lanterns. Wasps feed on insects, fruit and sugary juices. In late summer they can become a nuisance and bother people, because they want to feed on their lemonade or ice cream!

BUTTERFLIES are perhaps our most beautiful and colourful insects. The best time to examine them is when they are sunbathing on walls with their wings spread out. Often you will see two butterflies flying in rings around each other; this is a mating couple's courtship flight. They feed on nectar and use a long thin tube called a proboscis to suck it out of flowers. Nature has clever ways of protecting butterflies from predators, such as birds. As well as a means of attracting mates, the colour and design of a butterfly's wings warn birds that they taste very bad. Even butterflies that taste good to birds sometimes have wing colours and patterns to pretend that they taste awful! Many butterflies live for only a few weeks, but some can survive for a year.

Butterfly wings are very delicate, and so you shouldn't catch or hold them. Their wings are easy to damage, and a damaged wing will prevent a butterfly from flying.

The **SNAIL** is a creature with a long, moist and slimy body. Snails live in spiral-shaped shells, which they carry on their backs. If disturbed, they will withdraw into their shells and hide. They have four tentacles or feelers on their heads, and their eyes are on top of the two longest tentacles. Snails move along very slowly but they can live for up to ten years.

A **SLUG** is like a snail but it has no shell. Gardeners do not like slugs because, although they normally eat rotting vegetation, they also eat young plants. For this, they each have 27,000 teeth! They mostly live underground: every cubic metre of an average garden can contain over 100 slugs, and for every 5 slugs you find above ground there are 95 under the earth. Slugs seem even more slimy than snails, and their blood is green! They have lots of babies: one slug can have as many as 50,000 grandchildren! It's no wonder gardeners don't like them.

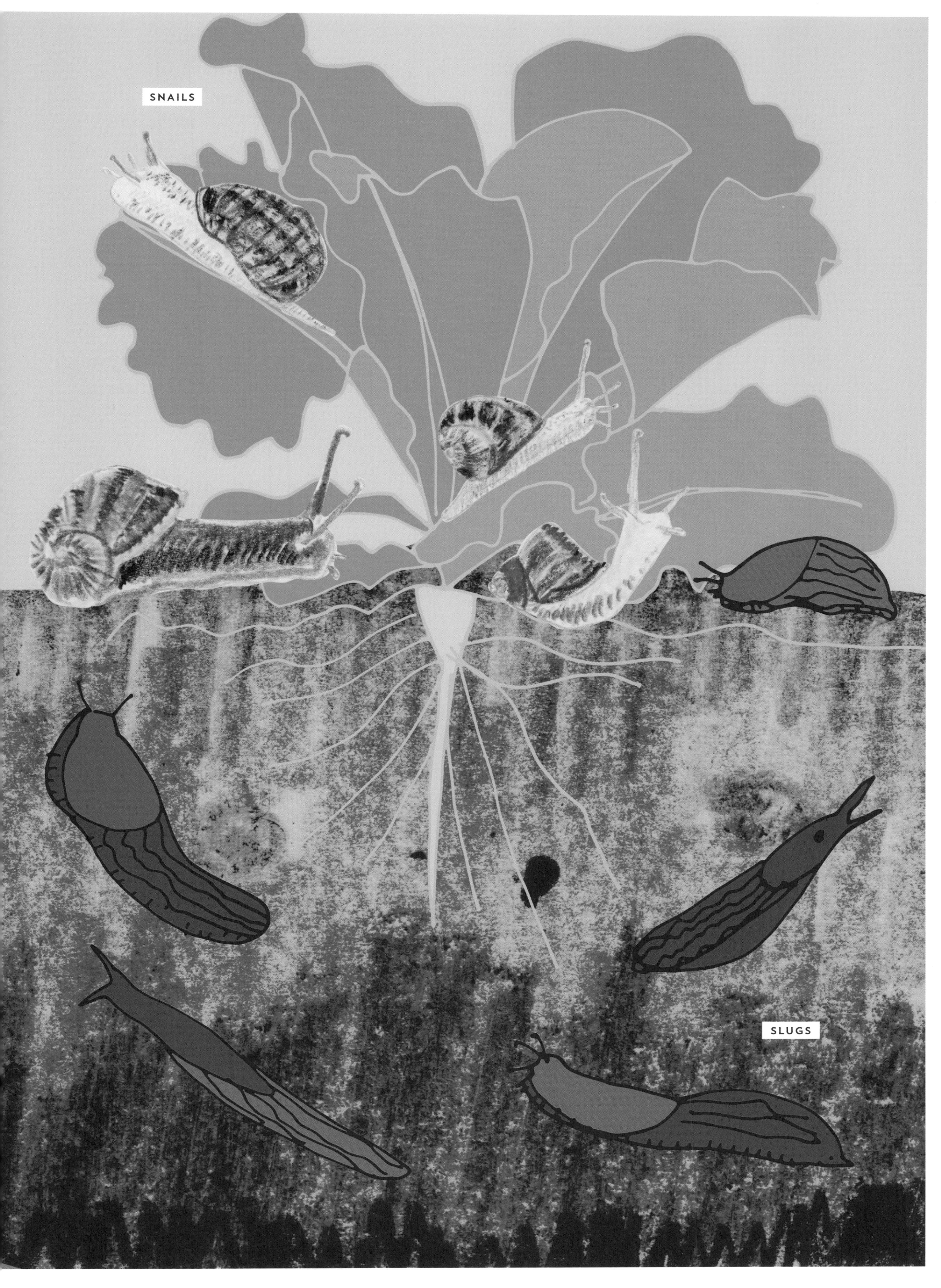
SNAILS
SLUGS

IN SUMMER THERE are not many birds to be seen in the garden because they are hidden in the thick foliage of trees and bushes, and are busy finding food to feed their young. Did you know that some birds have two or three families of chicks during a warm summer?

BLUE TITS can be seen flying back and forth to their nests every few minutes, bringing food to their chicks. A young family of blue tits needs to eat as many as 15,000 little caterpillars before they are ready to fly! It is important to feed the chicks well, because the more caterpillars they eat, the stronger the yellow colour will be in their feathers. When they grow up, this yellow colour is important because females always choose mates with the brightest breast feathers, a sign that the male will be a good caterpillar catcher!

The **BLACKBIRD** loves to search for food in the lawn. He hops along, his head cocked to one side, as he listens for a worm in the ground. Watch him closely; he must have excellent hearing because when he hears one, he digs fast with his beak and draws up a long worm and swallows it!

By late summer the garden is quiet and even the robin's song is missing. Do you know where the birds are gone? Well, every year they have to change their feathers and get a new fresh suit of clothes! While the old feathers are falling out and new ones are growing, the birds hide away quietly in the undergrowth.

BLACKBIRD

WHAT YOU CAN SEE IN

The Park in Summer

IN THE PARK trees are putting on a great burst of growth, both in height and in girth or thickness. The skin of a tree is called its bark. The bark is like a thick coat that protects the tree against disease and against some insects and animals, which could harm it. Trees increase their girth every year by adding a new layer of wood under the bark. When a tree is cut down, you can see that each year of its life appears as a ring. We can find out the age of a tree by counting the rings from the very middle to the outside. Some old trees in Irish parks can be as much as two hundred years old!

Sunlight is the most important thing for trees, and in a wood they have to race each other towards the sky to get as much sunlight as possible. Weaker species that do not win the race and are left behind do not get enough sunlight and eventually die.

Trees grow a lot in the late spring and the early part of summer. Then they begin to store food in preparation for the winter and to make new leaves when they start growing again the following spring.

WHAT YOU CAN SEE AT

The Seaside in Summer

SUMMER IS THE time when lots of people go to the seaside. It is a great place to explore because there are many wonderful creatures and plants to see along the seashore, between the land and the sea. The shore is covered by the sea twice a day, and when the tide goes out you can walk on what was, a short time before, the bottom of the sea. Only a special type of creature can survive this world that is wet half the time and dry the rest of the time. One is the **LIMPET**, a snail with a whitish cone-shaped shell. Another is the **PERIWINKLE**, a snail similar to the garden snail, but one that makes a nice shellfish dish when cooked. Some people like to pluck the periwinkle out of its shell with a pin and eat it raw! You can find limpets and periwinkles clinging onto rocks or in rock pools, waiting for the tide to come in again.

Make sure you have permission from a grown-up to explore the seashore.

ON BEACHES YOU can find many varieties of beautiful seashells that were once the homes of soft-bodied, snail-like marine creatures, protecting them as a crab's hard shell protects the crab. See how many different types of shell you can find. While you cannot easily see many of the creatures that live in the sea, when the tide goes out some of the smaller ones get trapped in rock pools. This is a good time to see these creatures as well as catch them in a fishing net. You can easily catch **SAND SHRIMPS** and small **CRABS** that usually hide under stones and seaweed.

Sand shrimps have ten walking legs and ten swimming legs. They are hard to catch in your hand because they can spring away very quickly to hide under seaweed.

Crabs come in all sizes (the Japanese spider crab can have legs that stretch as far as four metres, or twice the height of a normal door) but usually only small crabs are found in rock pools. They have an armoured body, eight legs with which they walk sideways, and a pair of sharp pincers which they use for catching their prey and signalling

BEADLET ANEMONE

to one another. Hermit crabs do not have a protective shell, so they find an empty sea shell, climb into it and carry it around with them. They escape into it when danger comes.

Attached to the rocks under water you will find a deep-red, jelly-like creature, about two centimetres in diameter, called a **BEADLET ANEMONE**. They feed by putting out their 192 tiny tentacles and catching sea creatures that are almost too small to see.

On some parts of the Irish coast you can see **SEALS**. These animals live in the sea and have four limbs that have flippers instead of hands and feet. They also have nostrils that they can close when they are swimming under water catching fish; some seals can remain under water for as long as half an hour. They can swim fast, but on land they are slow and awkward.

Another creature that can be seen in the sea in summertime is the JELLYFISH. In Irish these creatures are called SMUGAIRLE RÓIN, which means 'seal snot'! Some jellyfish can give you a bad sting, so do not touch them!

WHAT YOU CAN SEE IN

The Countryside in Summer

IN THE COUNTRYSIDE most livestock on farms live outdoors in the fields in summer, enjoying the warm weather and feeding on rich Irish grass. Towards the end of summer, farmers are busy in the fields, beginning to harvest crops such as **WHEAT, BARLEY AND OATS**, which were sowed in the spring or even as far back as the previous winter.

On many farms, besides wheat, oats and barley, you will see grass growing in fields. Grass is a very important crop for the farmer and some grass fields are used for cows and other animals to graze on. In other grass fields, the grass is cut to be stored as hay or silage to feed the animals through the winter.

On hill farms all the sheep are rounded up in early summer and have their thick winter fleeces sheared off with electric shears. The wool is sold and bought to make clothes and carpets.

...

MOTHS ARE SIMILAR to butterflies, but they fly mainly at night, and if you are in a car in the late evening you can see many in the headlights. They are attracted to bright lights, and sometimes on a summer's morning you may find some on the wall beside an outside light. In Ireland we have 33 different kinds of butterfly, but we have over 2,000 different moths! One moth, which flies by day and can be found in grassy areas near the sea, is the **SIX-SPOT BURNET MOTH**, which is black and red.

SIX-SPOT BURNET

The tiny **FROGHOPPER** is an insect that appears in summer and is a world champion jumper. It can be found in a little bubble of white froth attached to bushes and grass. Just six millimetres long, it can jump 70 centimetres into the air! The bubble of froth was called cuckoo spit in the old days, because people thought that it was left by the cuckoo. The froghopper produces the froth to hide in, away from birds or insects that might decide that he's a juicy meal. If you take a twig and probe the froth, you will see the little creature inside with its protruding frog-like eyes.

In the summer there are lots of birds to look out for in the countryside.

The **SKYLARK** is a small brown bird that is noticed only in the late spring and early summer. Twittering away and filling the air with its song, this bird flies so high in the sky – as much as 100 metres – that it is hard to see. When you do spot the bird, it is as if it is hanging there on an invisible string. The skylark's song continues without pause until the bird decides it needs a rest, and then it spirals down slowly, singing all the while.

SWIFTS come from southern Africa in the month of May to nest and have their young. They usually stay in Ireland for just two months before returning to Africa with their family. On warm summer evenings you can identify their pointed, crescent-shaped wings as they circle high in the sky, making a 'kee-kee-kee' squealing call. Except when they are nesting, swifts spend all their lives in the air, feeding, drinking, mating and even sleeping while flying, often four kilometres high in the air. Swifts like to nest in nooks and crannies in the roofs of old buildings. They fly amazingly long distances, often 800 kilometres in a day, catching flying insects to bring back to their young, sometimes a thousand at a time, which they keep in a big bulge in their throats. Swifts can fly at nearly 112 kilometres an hour, which makes them Ireland's fastest bird in level flight.

SWALLOWS

The other bird of summer is the **SWALLOW**, with its white breast and forked tail. It is a little smaller than the swift, and it makes its nest out of mud stuck high up on a wall in places like cowsheds. Swallows also fly from southern Africa to Ireland to have their babies. At the end of the summer the family flies all the way back to Africa.

One bird that is easy to identify in woods or parks is the **CHIFFCHAFF**. A small olive-green bird with off-white underparts, it is a summer visitor to Ireland, and sings a song just like its name, 'chiff-chaff-chiff-chaff-chiff-chaff'.

One of our animals is not often noticed by people although it is quite common, not only in the countryside, but in the skies above suburban gardens. The **BAT** is a wonderful little creature and looks a little like a mouse. It has wings stretched between its front and back legs. The bat is the only animal that can properly fly! In Irish it is called *an sciathán leathair*, or 'leather wings'. In the dark of night bats catch moths by using a skill called echolocation, like the sonar that warships use to detect submarines under the sea. Bats have amazingly sensitive ears and send out calls that we humans cannot hear. When this special call reaches a flying moth, it 'echoes' back to the bat. This way the bat knows where the moth is and can catch it.

Bats live in old buildings and roof spaces, often near a river or stream where there will be a lot of insects, moths or midges about after dark. You may catch a glimpse of bats if you are outside watching the sky getting dark on a summer's evening. After the swallows and swifts have gone home, but while the sky is still light, bats first appear. They do not fly as naturally or in as relaxed a manner as birds. Bats fly with fast, awkward and furious wing-beats, as if they are terrified that they will fall out of the sky if they stop! They dart about and change direction suddenly, chasing insects that you cannot see. Sometimes bats will patrol a bright outside light, because it attracts moths.

PIPISTRELLE BAT

There are eight different species of bat in Ireland. One of the most common is the **PIPISTRELLE**. It weighs no more than eight grams, or the same as a small coin, but it can catch 3,000 insects in one night's hunting. Many of the moths and insects they eat are harmful to farmers' food crops, so bats do a good job every night.

In winter, when there are few flying insects to feed on, bats find a warm cave or a hollow tree and they hibernate until the spring.

...

Ponds are very busy places in summertime. The **FROGS** have all gone away again, but by early summer the **TADPOLES** they left behind have grown bigger and are beginning to turn into little frogs, losing their fish-like tails and growing front and back legs. Soon they will climb out of the pond and look for a shady, damp place to live.

Some creatures live on the surface of ponds, including the shiny **WHIRLIGIG BEETLE**, numbers of which can be seen swimming fast in circles around one another. They have special eyes that allow them to look above and below the water at the same time, always searching for their prey of tiny insects. Whirligig beetles can carry a bubble of air that allows them to dive deep into the pond to seek food.

The **POND SKATER** is so light that it can walk on water using its long, slender legs, in such a way that it looks as if it is skating. Just like a spider is attracted by a fly in its web, the pond skater can detect the ripples made by an insect or fly that falls into the water. When it detects ripples, the pond skater races over and eats the unfortunate creature.

Another interesting insect that lives on the surface of ponds is the **WATER BOATMAN**. About ten millimetres long, it feeds on plant material and moves across the water paddling its long back legs like the oars in a rowing boat.

If you are lucky, you may spot a **NEWT**, a small lizard-like creature with a long tail that swims under the water. It is called an *earc luachra* in Irish, which means 'creature of the rushes'. Newts look like tiny dinosaurs. They are about seven to ten millimetres long and brown-green in colour, but during the mating season the male has a bright orange belly, and grows a wavy crest along his back and tail. Newts can live for up to eight years and, like frogs, they also leave their spawn in the pond after mating. If you do find newts, you must not disturb them. It is against the law to capture them!

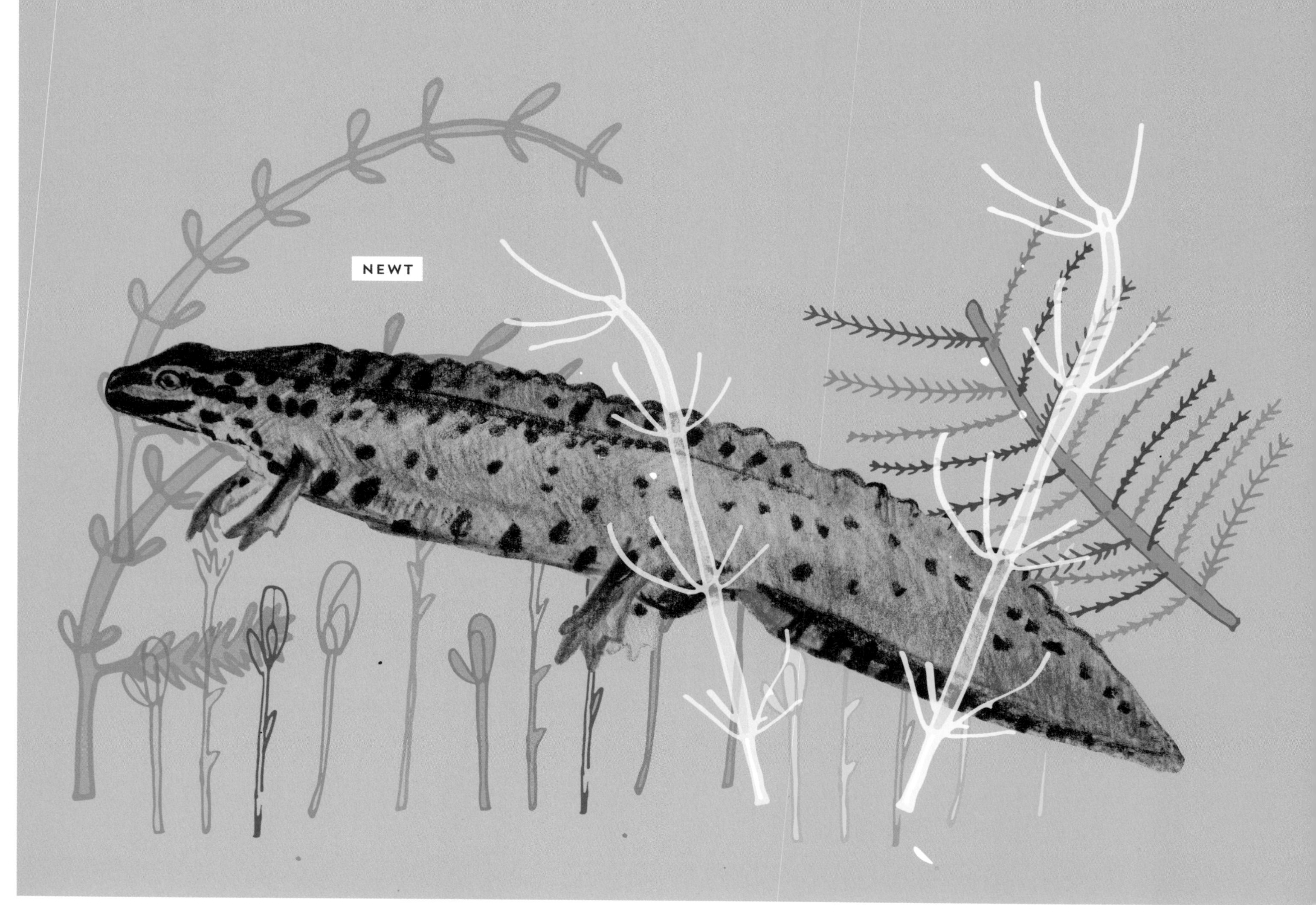

The **PYGMY SHREW** is Ireland's smallest animal. It is a tiny, mouse-like creature with a long nose. Pygmy shrews are always hungry and have to eat their own weight every day to survive, so they spend most of their time racing around and feeding on woodlice, spiders, beetles and other insects. Shrews like to live in hedgerows, which are full of insects. They are safe from predators when deep in the bushes.

The biggest wild animal in Ireland is the **DEER**, and our largest deer is the red deer. Deer can be seen mostly in mountainous areas, but the easiest herd to observe is in the Phoenix Park in Dublin. The male is called a stag, and stags grow antlers on their heads, just like Rudolf the Red-Nosed Reindeer. The antlers fall off each year and they grow bigger ones for the following year.

Another wild animal we have in Ireland is the **FOX**, a member of the dog family. A female fox is called a vixen, the male is called a 'dog fox' and a baby fox is called a cub. Foxes have beautiful red/brown fur and white bellies and throats. Long ago they were killed for their fur, which some ladies wore as a warm stole, a kind of scarf. The fox's home is called a 'den' or a 'lair' and is usually a hole in a ditch. In towns their den could be under a garden shed. Farmers do not like foxes because they can break into hen houses at night to kill and carry away their hens.

AUTUMN

AUTUMN IS THE SEASON WHEN ALL THE FOODS AND FRUITS THAT HAVE BEEN BASKING AND GROWING IN THE SUMMER SUN BECOME RIPE AND ARE READY TO BE GATHERED.

IN THE OLD days autumn in Ireland began at Lúnasa, the name of a feast day early in August when people celebrated the harvest – the gathering of all the food that would keep them going through the winter months when there was little food available. Today autumn officially begins on 1 September. The Equinox, when the day and the night are the same length and have 12 hours each, is in mid-September. From the Equinox on, the nights get longer and the days get shorter. Our part of the Earth is tilting away from the sun, and therefore it gets less sunshine every day. As the air cools down, the weather begins to get chilly, and we put on warmer clothes when we go out. For boys and girls the beginning of autumn means the end of the summer holidays, when they return to school to meet their school friends again, and look forward to the exciting feast of Halloween on 31 October.

One of the main signs that tells us autumn is beginning is when the leaves on the trees begin to change colour, from green to brown, and they start to fall to the ground in great crispy, golden heaps.

Isn't it lovely to kick through the heaps of dry, rust-coloured leaves that have dropped off the trees?

In autumn nature begins to prepare for winter. Let's see what we can look for in the garden, the parks and the countryside.

WHAT YOU CAN SEE IN

The Garden in Autumn

ON MOST PLANTS the flowers have withered by the autumn and they are now producing seeds. Try to collect seeds from flowers: if you carefully sow what you have collected in a pot, the seeds will produce flowers the following year. Are there any flowers still blooming in your garden in autumn?

IVY, which many gardeners do not like, is one of the few plants that happily flowers in autumn, producing blooms that are rich in nectar. Because there aren't many other flowers offering nectar at this time of the year, bees, hoverflies, wasps, flies and butterflies are very happy to visit the ivy.

The leaves of the **VIRGINIA CREEPER**, which is grown in many gardens to cover walls, are ablaze with many deep shades of crimson at this time of the year. Virginia creeper looks particularly beautiful when the setting sun shines on it.

MICHAELMAS DAISIES are popular in gardens because they, too, provide plenty of colour in the late summer and autumn.

...

WHEN THE WEATHER starts to get chilly, many insects are coming to the end of their lives, but they lay eggs in safe places to ensure that a new generation will be ready for the springtime.

SPIDERS of different kinds are common in the garden during the summer and autumn. Irish spiders usually live for two or three years. They lay from 100 to 2,000 eggs, and most baby spiders hatch in warm weather, but it is usually in autumn that we see signs of them. When they hatch, baby spiders spin a long length of almost invisible web to catch the breeze and take them flying away to a new place to make a home. On early autumn mornings, or in the afternoon when the sun is low, it is easier to see these flying webs.

The most common type of web to look for is the orbweb, which can best be seen on misty or frosty mornings. Orbwebs are like circular nets attached to twigs or branches to catch flying insects that are passing by. Orbwebs are among nature's most beautiful creations. In spite of having eight eyes, spiders cannot see very well, but are sensitive to any vibrations from their web. When a fly is caught in its web, the spider will immediately run across the web and bite it, paralysing it with venom. Spiders don't eat their prey, they inject it with chemicals that turn its insides into liquid, which they then drink! That's why you will see the remains of flies and insects hanging in a spider's web.

One of the smallest creatures you can see in the garden is the **RED SPIDER MITE**. It is not really a spider, but it looks like one. They are less than one millimetre in size and are best seen, often in large numbers, on white windowsills on sunny days. Gardeners do not like red spider mites because they feed on all kinds of plants, including tomatoes and strawberries, and damage them.

ANTS live in large colonies under the ground, often under the paving slabs on a garden path. In early autumn, on warm clear days, ants come up to the surface in large numbers to see off their queens, which have grown wings and are taking flight. The queens fly off in swarms, often attracting birds, like seagulls, which circle around the sky trying to catch and eat them. Those that survive lose their wings again as soon as they land far away from where they started, and they start new ant colonies. Ants eat a wide range of things from fruit to vegetables; some ant colonies even have their own farms! They encourage the growth of food, such as mushrooms, and they even have herds of greenflies because they like the honeydew that the greenflies produce!

The colourful **RED ADMIRAL** and **PEACOCK** butterflies can still be seen in the garden in September, but they, like lots of insects and animals, are beginning to look for cosy places to hibernate, or 'sleep over', for the winter, when there is not enough food for them. The peacock butterfly has big 'eyes' on its wings to scare off predators.

...

IN AUTUMN BIRDS that have been quiet during the summer start to sing again. Listen out for the different birds in your garden and see if you can name them from their song. In autumn and winter you can attract lots of birds into your garden by providing them with food.

The **ROBIN** is the easiest to identify with his orange/red breast and his spindly legs. His song is a tinkling melody, and his warning call sounds like 'tek-tek-tek'. The robin is a fierce defender of his territory.

The male **BLACKBIRD** is jet black with a bright yellow beak, while the female is a dull brown colour. The male blackbird has a wonderful song which he sings from the top of a bush, tree or rooftop.

The **BLUE TIT** has a blue cap and a yellow breast, and loves darting to and from a bird table.

His cousin, the **COAL TIT**, has a black head, white cheeks and a buff breast. He is smaller than the blue tit and not as brave, but he also likes to feed from a bird table.

Another cousin, the **GREAT TIT**, has white cheeks, a black head and bib over a yellow breast. He is the bully of the tits and chases other birds away from the bird table.

The **SPARROW** is brown and grey and is always chirping. Sparrows like to move around in small flocks.

COAL TITS

GREAT TIT
ROBIN

The male **CHAFFINCH** is a show-off with his plum-coloured face and breast, blue cap and brown wings with white stripes. His name in Irish is **AN RÍ RUA**, which means 'the red king'. The female is a dull brownish colour with a pale breast and white striped wings like the male. The male has a short but melodious song and a call that sounds like 'pink, pink, pink'.

The **SONG THRUSH** is about the same size as the blackbird, but is brown with a pale speckled breast. Like the blackbird, the song thrush has a beautiful warbling song.

The **WREN** is Ireland's second smallest bird (the smallest is the less common **GOLDCREST**). The wren makes up for its size by its loud and strident song. It is shy and usually moves around the garden under cover of bushes and shrubs, feeding on insects.

Do you have any of these birds in your garden?

WRENS

WHAT YOU CAN SEE IN

The Park in Autumn

THE PARK IS a place where you can see plenty of large trees. Autumn is a good time to identify different trees by their leaves, lots of which you will find on the ground.

Trees are wonderful natural machines: they help make our world beautiful, provide homes for creatures such as birds, insects and animals, and their leaves keep the air we breathe pure and healthy. Powered by the sun, trees suck up water from the ground through their roots, and through their leaves they take out of the air a harmful gas called carbon dioxide.

They turn the water and carbon dioxide into glucose, a kind of natural sugar, and into a healthy gas called oxygen. They use the glucose to make them bigger and

stronger, and they let out the oxygen into our air, making it clean and healthy for us to breathe. This process is called photosynthesis. From autumn to winter, there is not enough sunlight to operate the great tree machine, and they know it is time to take a long sleep, during which they will quietly get ready for spring. During their winter sleep, trees live on the food they have been storing up since the previous spring.

In autumn some of the glucose and other minerals that the tree does not need become trapped in the leaves and turns them from green to beautiful shades of orange and gold. When trees have finished using their leaves, they let them fall to the ground, where they become food for lots of worms and insects. This way they eventually pass their valuable minerals back into the soil. Trees also produce their seeds in autumn in the form of berries and nuts, which, if planted, will eventually become young trees.

In autumn we can see how nature gets creatures and plants big and small to work together and look after one another. The great big tree feeds the tiny insects and worms on the forest floor with its leaves, and they in turn enrich the soil, which helps the tree to grow bigger the next year. See if you can find **EARWIGS**, **BEETLES**, **WORMS** and **SLUGS** under the crispy carpet of leaves on the ground.

Trees also feed birds with their berries and nuts, and birds spread these seeds to create more trees!

Here are some trees to look out for.

The **OAK** is called 'The King of the Woods'. Ireland's oldest oak is said to be over 400 years old and grows in the grounds of Charleville Castle in Tullamore, County Offaly. Oak is a very strong timber and was once used to build sailing ships and castles. In ancient times doors that were made from oak were said to keep out evil spirits! Oaks produce shiny green acorns in little cups, which are their seeds.

The **ASH** is the most common tree in Ireland and has lots of small leaves on each stalk. The tallest ash in Ireland is in Clonmel, County Tipperary; it is 40 metres high! The seeds of the ash, which hang from its branches like a bunch of keys, provide food for birds, mice and squirrels.

The **BEECH** is a tall, grey-barked tree with small leaves, delicate green in spring and golden brown in autumn. Its seeds are shiny triangular nuts in a rough brown case: they once were used to feed pigs.

The **SYCAMORE** is the second most common hedgerow tree in Ireland. Its seeds are like tiny helicopters, which, when ripe, spin away on the wind. This is how the sycamore spreads its seeds far and wide. Collect some of the little helicopters and see if you can make them fly.

The **HORSE CHESTNUT** produces leaves that are like five long, thick fingers, and shiny brown nuts in prickly green cases. Have you ever played conkers with the nuts?

Can you tell which tree is the last to lose its leaves in autumn?

In the park you will spot bigger birds which you might not see in your garden, such as **WOOD PIGEONS**, **ROOKS** and **MAGPIES**. If there is a pond in the park, you will see graceful **SWANS** and shyer birds, such as the **GREY HERON**. The grey heron is a large bird with long legs and a yellow dagger-like beak for spearing small fish.

WOOD PIGEONS are fat, grey birds with little heads and white patches on their shoulders. They can usually be seen grazing in grassy areas, eating grass seed. Wood pigeons do not have a pleasant singing voice, but have a distinctive hoarse call that sounds like 'coo-coo, cuck coo, coo', which they repeat over and over.

ROOKS are large members of the crow family and are sheeny black with a grey beak. They feed on creatures called grubs, which they dig up from grassy areas with their big beaks.

MAGPIES are the 'bad kids on the block', easily identified by their distinctive black and white colouring. They go around in gangs and are always bullying other birds and animals. In spring magpies often steal eggs from a smaller bird's nest and eat them!

The SWAN is one of the world's heaviest birds, but is beautifully graceful with a long curving neck and an orange bill or beak. Swans can weigh up to 18 kilograms, which means they have to work hard to get into the air when they start to fly. In the air swans are even more graceful than on water, and their great wings make an unusual throbbing sound. A young swan is called a cygnet. You should never go too near their cygnets because swans can get very angry, particularly when they are on their nest protecting their young! Swans have a special place in myths and legends, including the old Irish legend of the 'Children of Lir'. The legend is about the children Fionnuala, Aed, Fiachra and Conn being turned into swans by their jealous stepmother, Aoife.

The HERON is a tall bird with a grey and black body and a white neck. Its long legs allow it to walk in streams and ponds, stalking small fish. Sometimes the heron stands absolutely still, looking down into the water, and then suddenly it strikes, spearing a small fish with its beak and swallowing it whole. It is often chased by flocks of smaller birds, and it flies away with long, slow wing flaps, protesting with a harsh 'snark!' call.

GREY HERON

WHAT YOU CAN SEE IN

The Countryside in Autumn

LATE SUMMER AND autumn are busy times for farmers. Crops such as **WHEAT, BARLEY AND OATS**, which have been sown in the fields in spring or early summer, are ripe now and they must be harvested or collected before too much rain damages them. It is exciting to see the harvest under way. Very large and noisy machines called combine harvesters drive through fields cutting the crop, separating the grain from the stalks and pouring it into tractor-drawn trailers. Wheat grain is ground into flour, which is used in baking bread. Barley grain is an ingredient in various foods, medicines and beer. Oats are used to make porridge, which has been a favourite Irish breakfast dish for hundreds of years.

Soon after the crops have been harvested, the land has to be ploughed again so that new crops can be sown for the following year.

The fields in the countryside are full of beef cattle in autumn. The cattle have been fattened up on healthy grass over the summer and are now ready to be sent to the market to be sold. Autumn is also the time when fruits in the orchards, including **APPLES, PLUMS AND PEARS**, are ready to be picked and packed and sent to market.

Autumn is the time when we see lots of mushrooms of all shapes, sizes and colours. Mushrooms are the flowers of fungi that live under the ground. While the common field mushroom is edible, most mushrooms are not suitable to eat.

Many mushrooms are very poisonous and should not even be touched! One poisonous but pretty mushroom is the **FLY AGARIC**, the typical fairy mushroom with a bright red top covered with white speckles. You will also find mushrooms on trees and dead logs; these are part of nature's recycling system, turning dead wood back into soil.

FLY AGARIC

Many delicious wild fruits also become ripe in the autumn. Country hedges are ablaze with colourful berries, including some that we can eat and use in cooking, like **BLACKBERRIES** and **CRAB APPLES**.

Blackberries, which grow on the bramble bush, are the easiest to pick. They can be made into delicious jams and tarts. In the old days they were mashed up and eaten with porridge, and the leaves of the bramble that they grew on were used to treat sick farm animals.

Crab apple trees grow in hedgerows all over Ireland. Their little apples were a valuable food source in ancient times in Ireland, and are mentioned in many old stories. Crab apples are much smaller than cultivated apples, and a little too sour to eat, but they can be used to make delicious crab apple jelly.

In the early autumn little purple berries called fraughans or bilberries become ripe on mountainsides. They are good in pies and are delicious eaten directly off the bush, although they may give you a purple mouth!

The **HAZELNUT** grows on hazel trees, which can be found in hedgerows and in woodlands. In the old stories the tree was important because the hazelnut was said to contain knowledge and wisdom. Hazel branches were used to make the walls of houses in the Stone Age. The nuts are good to eat, either straight from the shell, if ripe, or cooked in a hot pan. They are also an important food source for mice, shrews, pigeons and pheasants.

Many hedgerow shrubs, like the **BLACKTHORN** and the **HOLLY TREE**, produce berry harvests for birds and help them fatten up for the winter. These shrubs make sure to produce brightly coloured berries to attract birds because, when birds eat them, the seeds in the berries are dropped in the birds' poo and a new shrub grows! Some berries, such as those from the holly and the yew tree, are poisonous to humans, so never eat wild berries without asking an adult first.

...

IN THE COUNTRYSIDE some animals, including the **HEDGEHOG** and the **BAT**, spend the autumn getting ready to hibernate, which means they go to sleep for the winter.

BATS are small and find it hard to keep warm in the cold winter weather, when their usual diet of flying insects and moths are not plentiful. Instead, bats fill themselves up with food during the autumn and then find a sheltered, warm place, such as the attic of an old house or a hollow tree or a cave, to sleep. When they find the right spot, a number of them will hang upside down and close together for warmth, with their wings folded around themselves. There they will stay until spring comes.

HEDGEHOGS also spend autumn feasting on food to build up their fat layers to last them through the winter. Then they find a cosy spot in a bundle of straw or dry leaves, make a nest for themselves and snuggle down for a long sleep until spring.

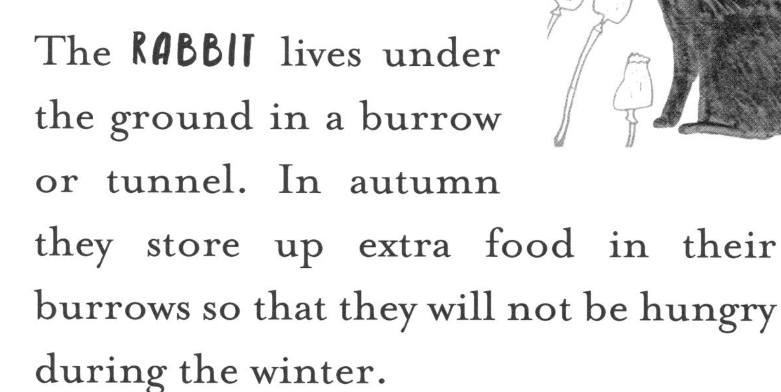

The **RABBIT** lives under the ground in a burrow or tunnel. In autumn they store up extra food in their burrows so that they will not be hungry during the winter.

In the west of Ireland you may be lucky enough to see the **RED SQUIRREL**, which is smaller and much shyer than the grey squirrel. It eats nuts and berries, but its favourite lunch is oil-rich pine seeds. The red squirrel spends most of its time foraging or collecting food high up in trees, compared to the grey squirrel, which forages a lot on the ground. When you approach a red squirrel in a tree, it will hide behind the trunk, occasionally taking a peep to see if you are still there. If you stay absolutely still, it often thinks you are gone away and will resume its search for food!

The **BADGER**, with its grey body and black and white striped face, is one of our biggest wild animals. It does not hibernate but eats a lot and gets fat in the autumn to be ready for the winter, when food is scarce. Badgers spend most of the winter resting and saving energy in their burrows in the ground. Badgers are not easy to see because they are nocturnal creatures – they sleep during the day and mainly come out at night.

In the autumn the male **DEER**, the stag, will fight other stags to win a mate. It is a noisy crashing affair, while they run at each other with their heads down, clashing their antlers together.

IN THE COUNTRYSIDE you will see all the birds that frequent gardens and parks, but in the autumn you may also get a chance to spot some of the bigger migratory birds – birds that do not stay in Ireland all year round, but come and go once a year.

The winter in Ireland is much warmer than it is in many northern countries. Because of this, hundreds of thousands of birds arrive here from the north during the autumn and early winter for their holidays. They include wild **GEESE** and **DUCKS** and two kinds of thrush – the **FIELDFARE** and the **REDWING**. Some of them travel great distances from Greenland and Iceland where, in winter, the ground is frozen hard and birds cannot get worms to eat. The ground in Ireland doesn't often freeze hard, so you can see great flocks of these visitors feeding in our grassy fields and parks. Geese fly in big V-shaped formations, which are wonderful to see. The smaller birds, such as fieldfares and redwings, often turn up in gardens to feed on berries, and you may spot them in your garden if you're lucky.

Other birds that have been with us all summer, such as the **SWALLOW** and the **SWIFT**, prepare in autumn to leave Ireland and fly south.

Swallows spend their days flying and feeding on flying insects. Sometimes you can see them fly low over ponds and dip their beaks in for a sip of water. Swallows have to leave Ireland in the autumn because there aren't enough insects to feed on here during the winter. They fly all the way to southern Africa, nearly 10,000 kilometres away, where it is summertime and there is plenty of food for them. As they get ready to go, swallows fly round and round in family groups, twittering all the while. The parent birds are preparing their babies for the long flight south. The young swallows are just three months old, and only about one in three of them survive the journey. Before they fly away, swallows gather in large numbers on telegraph wires and roofs – see if you can spot them just before they leave. When they return the next year, they always come back to the same nest they used the previous year.

SWIFTS are bigger than swallows, and also spend the day feeding on flying insects. At night they fly very high, and they sleep while they are flying. Imagine that! They arrive in Ireland later than swallows and leave earlier, again flying south with their young to southern Africa.

In autumn the **STARLING** changes from its summer plumage into a new speckled winter coat. Many more starlings, flying from the cold north, come to join the birds that have been with us in Ireland during the summer. Starlings like to fly about with their friends and families in flocks. Sometimes in autumn and early winter they can be seen in flocks of many thousands. Starlings are excellent mimics and like to copy the songs of other birds. Some have even been heard copying car alarms and chainsaws!

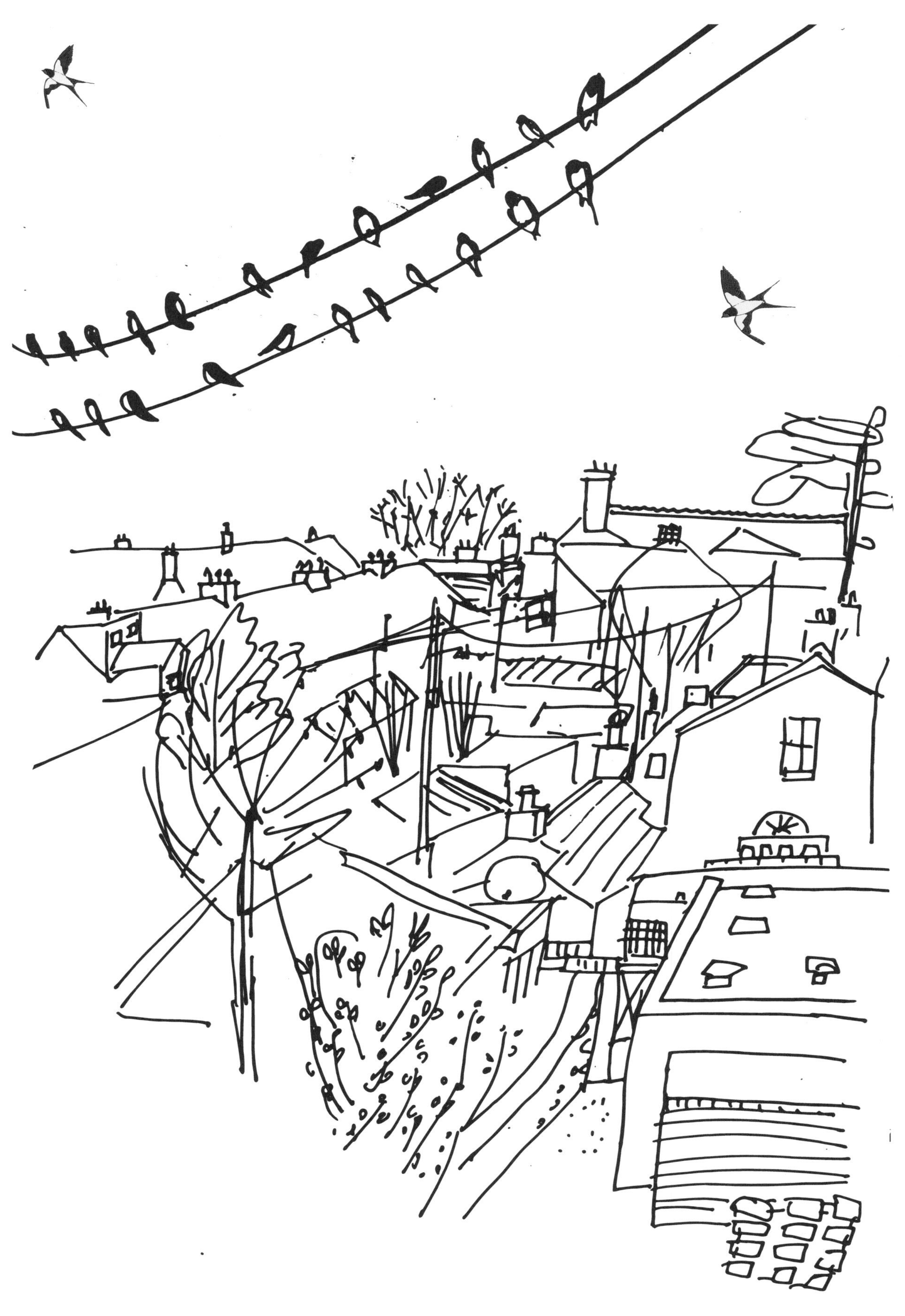

WINTER

... ...

WINTER IS THE NIGHT-TIME OF THE YEAR. DURING WINTER, NATURE RESTS, SPENDING MUCH OF IT SLEEPING AND GROWING STRONG AGAIN IN PREPARATION FOR THE COMING SPRING. IT IS A HARD TIME FOR ANIMALS.

...

FROM THE BEGINNING of winter, children look forward to Christmas, which comes a few days after the shortest day of the year, 21 December. On the shortest day, our part of the Earth, called the northern hemisphere, is tilted the farthest away from the sun. While it is winter in Ireland, people in places like Australia, in the southern hemisphere of the Earth, enjoy their summer: the longest day of their year is 21 December.

In Ireland, after 21 December, the days begin to get a tiny bit longer every day and nights get a little shorter.

Some creatures, such as hedgehogs, shrews, field mice and bats, can

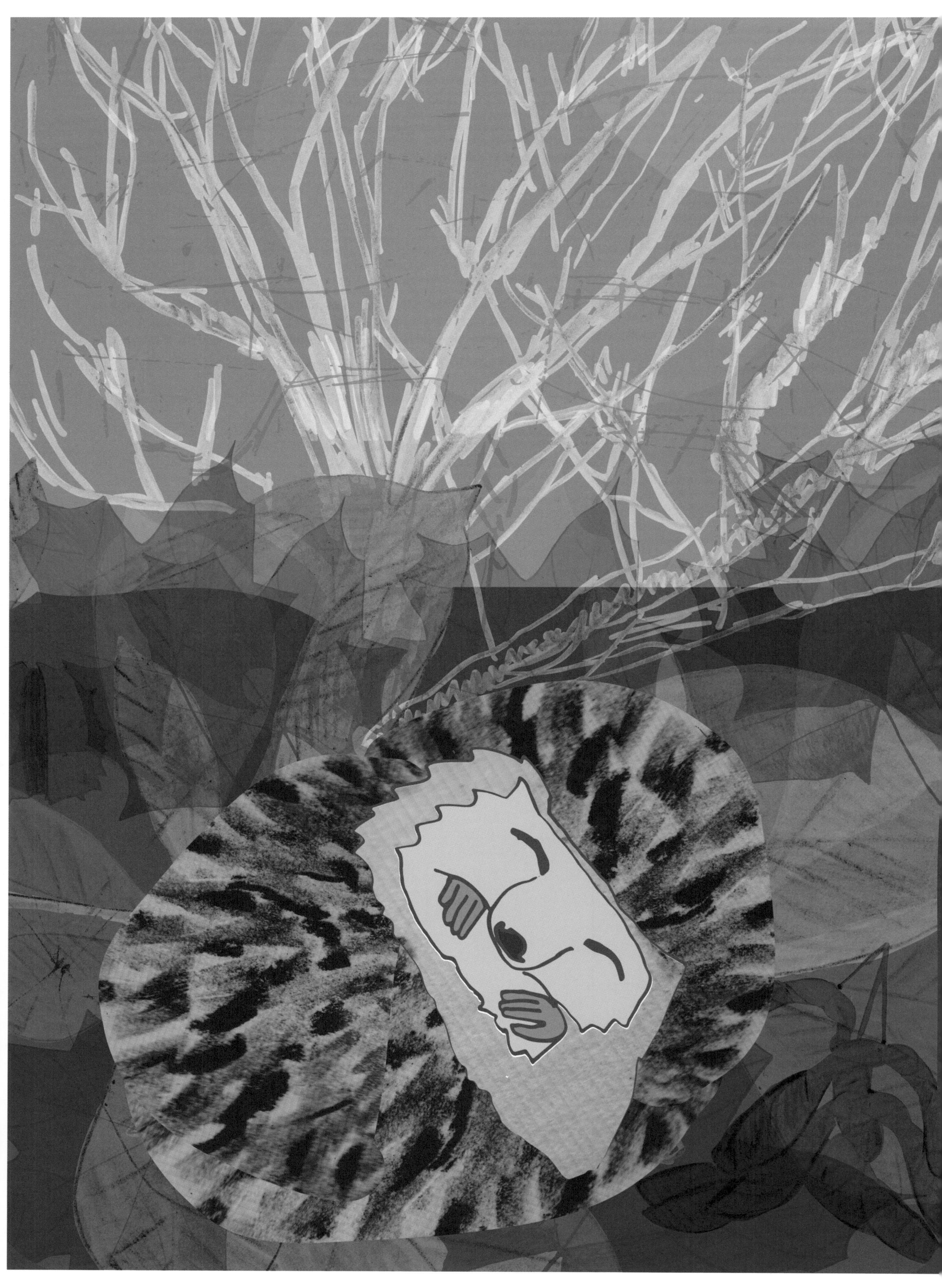

slow down their heart beat and lower their temperature when winter comes. They do this so that they can quietly sleep through the cold months in the cosy nests they have prepared: these animals are said to hibernate. Many of our insects, including the small tortoiseshell butterfly, the peacock butterfly, and queen bees and queen wasps, also hibernate in winter.

By the time winter comes, some of our birds have already flown south to spend a few months in warmer countries. Other birds, from northern regions, come to Ireland for their winter holidays because it is warmer here than where they spend the rest of the year.

Winter is a good time to study nature and, perhaps, start a nature diary in which you can note all the interesting things you can see in the garden, the park and the countryside. Without the many leaves of summer, it is easier in winter to examine trees. You can see how their branches are arranged and how each type of tree has different kinds of branches. It is also easier without the leaves to spot birds and identify them in winter. When you hear a bird singing, listen carefully to the song and try to see what kind of bird it is. The next time you hear that song, you will know without seeing it whether it is a robin or a wren. Winter is also a good time to find abandoned birds' nests in shrubs and bushes and see how they are made.

So let's put on our warm clothes and go out to see what we can find in the garden, the park and the countryside.

WHAT YOU CAN SEE IN

The Garden in Winter

IN WINTER THERE are no butterflies flitting through the air, no hum of bees, and nearly all the flowers and blossoms seem to have died. There are, however, a few plants that thrive at this time of year. See if there are any in your garden.

WINTER JASMINE is a garden creeper that seems to think that the summer has arrived and puts out lots of little yellow, perfumed, star-like flowers.

Garden shrubs such as the **PYRACANTHA** and the **COTONEASTER** produce lots of red/orange berries in winter, which hungry birds like to feed on. If neither shrub is in your garden, see if you can spot a pyracantha or a cotoneaster in your neighbourhood.

Nature makes sure, however, to provide some wild plants that flower and produce berries in the winter. **IVY** is a creeping plant that gardeners don't like, but there are good reasons for having a little wildness in the garden. Ivy is an important plant because it produces autumn flowers that provide high-quality nectar for bees and insects when most other flowers have withered. The berries feed wood pigeons, thrushes and blackbirds through November and December. Ivy also provides good nesting places for birds and insects, and, during winter, hibernating places for butterflies.

IVY

The **SNOWDROP**, just as its name suggests, is like a little drop of pearly white snow on a tiny stalk between two long narrow leaves. It pops out of the ground in late winter and frost doesn't seem to do it any harm. The earliest the snowdrop has been found in Ireland is 2 January. It is not a native plant; our snowdrops probably came from Spain and France hundreds of years ago.

GREENFINCH
COTONEASTER

YOU WILL NOT find many insects in the garden in winter. While most animals have body fat and generate their own heat to protect themselves from the cold, insects mostly have to rely on the warmth of the sun, and there isn't much warm sun in winter. Some insects last only for the summer, but lay eggs that will produce new young insects the following spring. You may find some of these tiny eggs under logs or stones. Some insects with wings, such as beetles and butterflies, can fly away to warmer places, just like birds. The most famous migrating butterfly, the monarch, flies over 33,000 kilometres at the beginning of winter from the cold of Canada to the warmth of Mexico. Other insects survive by hibernating in a dry, warm place where the frost cannot get at them. Some even have special body fluids that keep them from freezing. These special fluids are like antifreeze, which people put on car windscreens early on winter mornings to remove the frost!

Ants delve down into the ground, below where frost will reach, and there they will feast through the winter on the food they have stored up during the summer months.

...

THE BIRDS THAT don't fly away to a warm country have to eat a lot to stay warm. You can help them and bring colourful activity to your garden by providing regular food for them on a bird table.

A bird table is just a flat piece of wood on which you can lay out tasty snacks. Get help from an adult to put it up. It should be placed where you can easily see it from a window. It is best if the bird table is put high on a post or a branch, so that cats cannot sneak up on the birds when they are feeding! If you have grey squirrels in your area you may also have to find a way to prevent them from stealing the food you put out. It is good to have your bird table near a tree or a bush so that birds can easily and safely queue up to get food.

Peanuts are the most popular food to put out for garden birds in winter. Peanuts will attract **BLUE TITS**, **COAL TITS** and **GREAT TITS**, as well as less common birds such as **GREENFINCHES** and **SISKINS**. Peanuts should be held in a feeder that allows the birds to take only little pieces at a time, because they might choke on a whole nut. Make sure not to let your peanuts get mouldy, because they can then be very harmful for the birds. Peanuts should be changed every two weeks if they are not eaten.

Black sunflower seeds will attract seed eaters such as **CHAFFINCHES** and, if you are lucky, the beautiful colourful **GOLDFINCH**. Fat can be good for birds in winter. It can be either in the form of suet from the kitchen, hung off your bird table, or fat balls that can be bought in the supermarket.

BLACKBIRDS and **THRUSHES** like apples that are left out for them. Lots of waste food from the kitchen is suitable: bacon rinds and melon seeds are enjoyed by most birds. It is not good to feed birds with dry bread because it can expand in their throats and they could die.

Remember, once you start feeding birds, they will expect to be fed and will come to depend on you, so you should continue to feed them until the middle of spring.

A shallow container or bowl will provide a bird bath for the birds. It is fun to watch them drink the water and then splash about washing their feathers. The water in a bird bath should be changed regularly.

One of the birds you will see and hear a lot in winter is the bully of the bird world, the **MAGPIE**. It is easy to identify with its show-off black and

white coat and long black tail feathers. Magpies are always up to mischief, whether annoying other birds such as woodpigeons or crows, or ganging up on cats. They happily enjoy eating lots of different foods, from insects to other birds' eggs, and often, at nesting time, young nestlings!

The favourite Irish bird of the winter is the red-breasted **ROBIN**, pictured on many Christmas cards with snow and holly berries. Robins love to watch someone working or digging in the garden because it means insects that have been hiding in the clay or under leaves have to run away to escape. The robin is quick to spot them, and will dart down to make a meal of insects scuttling for cover.

The robin you see in the garden might not, however, be the one you saw during the summer or autumn. Some of our robins fly south to a warmer place in winter, and other robins from Scotland and Scandinavia come south into our gardens at the same time and take their place.

Another bird that you may not see in the garden in winter, but you will definitely hear, is the **WREN**. It is one of our smallest birds, but it has the

loudest and most strident call of all the garden birds. The wren lives in a twilight world in the thickest hedges and undergrowth, and has to sing loudly to be heard by other wrens. In the old days, groups of boys used to catch a wren and bring it in a cage around houses on St Stephen's Day, 26 December, singing a song and collecting money:

The wren, the wren, the king of all birds,
On Stephen's Day was caught in the furze;
Up with the kettle and down with the pan,
Pray give us a penny to bury the wren.

Once upon a time all the birds wanted to see who could fly the highest in the sky, for that bird must surely be the king of all birds. They all took off and climbed as high as they could. The eagle, of course, flew the highest. But just as the eagle became tired and couldn't rise any more, a little wren that had been hiding in his feathers came out and flew even higher than the eagle, and that's how he was crowned the king of all birds!

To fight the cold of winter, wrens often gather together, with up to two dozen of the little birds squeezing into a dry space in a stone wall to keep themselves warm.

...

THE LONG DARK nights of winter are good times to observe the night sky in clear weather. If you are in the country where there are not many lights, you will see the sky at its best, but even in the city, from a dark part of the garden, there is a lot to see. Besides the moon and the millions of sparkling stars that fill the sky, you can also see spacecraft and satellites passing over, and if you are lucky, you might see shooting stars.

The best way to look at the night sky in winter is to get together with your friends or family on a night without clouds, wrap up in warm clothes, and lie down on a mattress or on cushions in the darkest part of the garden. This way you won't get a pain in your neck and you will be able to examine the stars and the patterns they make. Long ago ancient people imagined they could see figures or pictures outlined by stars, and they made up stories about them and gave them names that are still used today. One of the easiest of these to spot is what we call 'The Plough'. This is a shape made up of seven stars roughly in the shape of a saucepan. Two of the stars of the Plough point to a star called the Pole Star, which always indicates which direction north is.

If you are patient, and if you watch at particular times, you will see shooting stars. Shooting stars are really meteorites or lumps of rock flying through space at very fast speeds. When they hit the layer of air that surrounds our world, they burn up and appear as sudden, brief streaks of light.

The other star-like objects you will see in the sky are Earth's satellites and spacecraft that are constantly going round and round the Earth. They look like stars, but you will see that they move steadily across the sky, while the stars seem to stay still. The first satellite was sent into space in 1957. Today there are more than a thousand working satellites circling the Earth, including a huge telescope called Hubble, which is used for studying the universe, and the International Space Station, which usually has a crew of astronauts on board. There are also more than 10,000 other large pieces of old spacecraft and used rockets flying around, so if you are patient, you are bound to see some of these passing across the sky.

The moon is the easiest thing to examine in the night sky. It is 384,400 kilometres from Earth and about 4.5 billion years old. Because it has no layer of air to protect it, as the Earth has, it has been hit by many meteors and meteorites. If you look at the moon through a telescope or binoculars, you will see that the surface is covered with craters caused by these great crashes. The moon is of course round, but it seems to change shape as the sun lights different parts of it. The darker areas on the moon, which you can see without binoculars, are expanses of lava that have come from moon volcanoes. In the old days, before telescopes, these were thought to be oceans and seas like we have on Earth, and they were given names such as the 'Sea of Serenity' and the 'Sea of Cold'.

The first man to walk on the moon was Neil Armstrong, who climbed out of his spacecraft onto the surface after a long spaceflight on 20 July 1969.

WHAT YOU CAN SEE IN

The Park in Winter

In winter ROOKS often gather on playing fields in the park to feed on worms. The best time to see large flocks is early in the morning. In some coastal counties they are joined by SEAGULLS and OYSTERCATCHERS, who are also fond of worms and have come all the way from the seaside.

There are other smaller winter visitors that you might see in your garden or in the parks in cold weather, such as two kinds of thrush – the redwing and the fieldfare. The REDWING is like an ordinary thrush but with rust-coloured markings just under its wings. The FIELDFARE has a grey-brown head, neck and back, and a light coloured speckled breast. Both the redwing and the fieldfare come to Ireland from Scandinavia.

...

ONE OF THE few plants in bloom in winter in parks is the **WINTER HELIOTROPE**, which can be found carpeting embankments and shady places under trees. Its leaves are round and almost heart-shaped, and it produces pink-white flowers from November to March. The winter heliotrope produces a strong sweet scent, and it is a valuable source of nectar for the insects that are still about in winter. Winter heliotrope is often planted near bee hives to provide winter nutrition for the few bees that leave the hive. The plant turns its flowers towards the sun and follows it through the sky to get as much solar energy as possible; at night it turns towards the east to await the sunrise. Other flowers that follow the sun like this are the **DAISY**, the **MARIGOLD** and the **POPPY**.

Some trees don't lose their leaves in winter. These are called evergreens. Among the evergreen trees that thrive in winter is the **HOLLY** tree. In December the rich, shiny green of its prickly leaves and its brilliant bunches of red berries

stand out against the drab winter shades of other trees and bushes. For this reason, holly is popular as a decoration at Christmastime. The Romans thought that a holly tree planted near a house defended it against witchcraft and being struck by lightning! The holly is mentioned in many of the ancient Irish legends, and magic spears are said to have been made from its hard wood. In the old days the holly was called *an crann uasal*, or the noble tree. It was thought to be protected by fairies, and terrible things would happen to anyone who cut down a holly tree.

The **SCOTS PINE** is tall and straight and also has prickly spines called needles. It has small flowers in the month of May, which turn into pine cones that provide a protective shell for its seeds. Red squirrels like to feed on the pine seeds, but have to break the cone apart to get at them. When you see the remains of cones on the ground under trees, you know there is a red squirrel about.

SCOTS PINE

GREY SQUIRREL

The **YEW** is another evergreen tree, so it doesn't lose its very thin leaves, which are called needles, in the autumn. All parts of the yew, including its red berries, are poisonous. The yew is never grown on farms because cows and horses might nibble it and die! The oldest yew tree in Ireland is said to be one that grows in Muckross Friary in Killarney: it might be nearly 700 years old!

By winter the **ASH** tree has shed all its leaves, but clusters of seed pods, called ash-keys, still hang from its branches.

In late winter the buds on trees begin to swell as the leaves inside them are starting to develop. Can you see which buds swell the most and guess which tree is going to produce leaves first?

The lakes and ponds in parks attract large flocks of ducks, geese and waders in winter. Waders are birds with long legs that allow them to walk in the water at the edge of ponds and lakes without getting their feathers wet. They move along slowly, probing the mud on the bottom of the pond in search of insects, tiny crabs and worms.

...

THE ANIMAL YOU are most likely to spot in the woodland part of a park is the **GREY SQUIRREL**. The squirrel is a member of the rodent family, like mice and rats. Rodents have sharp front teeth that are constantly growing, so they have to keep gnawing on food to prevent them from getting too long. All the grey squirrels in Ireland are descendants of a few pairs that were brought from the United States in 1911. They eat acorns, beech nuts, hazelnuts and pine seeds, and strip the bark from trees in autumn to get at the sap. During the breeding season they can often be seen chasing each other and wrestling. Squirrels bury nuts and pine cones in the ground in different places and come back to find them weeks later!

WHAT YOU CAN SEE IN

The Countryside in Winter

BY MID-WINTER most plants and flowers are withered and dried up, and their leaves and blossoms have turned golden brown or yellow, giving the landscape a different colour.

When the weather gets cold, farmers bring some of their animals into the warmth of barns and sheds. There they are fed throughout the winter on a food called silage, which the farmer makes from grass that has been cut in the summer. Some hardy animals stay outside in the fields for the winter, and because there is very little grass to eat, the farmer brings them food every day.

Some sheep are hardy enough to be left out on the hills, but the shepherd has to bring extra food up to them.

Herds of deer that live most of the year in the hills and mountains often move down to warmer places in winter and spend a lot of time in the cosiness of woods.

Along the coast, look out for **SEALS**. During November and December new seal pups are born with thick, creamy white fur to keep them warm. They keep this 'baby fur' only for a month or so, and then grow fawn-coloured adult fur.

...

Great numbers of **GEESE** and **WHOOPER SWANS**, who live the rest of the year in the Arctic regions of Canada, Greenland and northern Russia, fly south when it gets really cold there in winter. Our winter is like summer to them!

The wetlands and mudflats of Ireland, places where these birds like to feed, don't freeze over in winter, and places like this provide plenty of food for the visitors. Waders and geese like to gather on mudflats when the tide has gone out, feeding on the worms and other creatures that live in the mud. Many

waders have sensitive tips to their beaks which help them to find their prey in the mud. Different creatures live at different depths in the mud. The large speckled **CURLEW** has a long, down-curving bill that is 15 centimetres long, which allows the bird to get down deep enough in the mud to capture big fat **LUGWORMS** and **RAGWORMS**. The little **DUNLIN** has a bill less than three centimetres long, which reaches small creatures only just under the surface of the mud.

Some of these birds stay only for a while in Ireland, for a rest on their way south. Others, such as the 40,000 **BRENT GEESE** that fly here from Arctic Canada – nearly 3,200 kilometres away – and the 7,000 **WHOOPER SWANS** that come from Iceland, enjoy their stay until March or April. You can spot these winter visitors all around our coasts, but especially in the Wexford Slobs and places like Bull Island in Dublin Bay.

On large ponds or lakes, flocks of a bird called the **COOT** can be seen in winter. It is dark grey, nearly black, and has a white beak and white forehead. The coot feeds on almost everything you will find in a lake, including insects, seeds and plants. It will sometimes even steal food from other birds!

Winter is a time when you can see great flocks of **ROOKS** and **STARLINGS** gathering in the air, as if for a party, shortly before dusk. Rooks often assemble in a field near their roosting trees. Then they take off and perform an intricate series of aerial exercises, swirling and pitching about, speeding up and slowing down, cawing loudly all the while. Eventually, as darkness falls, the rooks plummet down to their roosts, and suddenly all is silent as they settle in for the night.

If you are lucky, you might see in the country a similar, but much more spectacular, gathering of starlings on a winter evening. It is probably the most amazing wildlife experience you can see in Ireland. This gathering of starlings is called a murmuration. The starling is a bird about the same size as a blackbird, and looks a bit like one, except it has white speckles and sheeny feathers. Starlings like to fly about and feed together in a flock, and they begin their murmuration activity in November. The numbers in their flocks increase as winter goes on. Just before dark in the evening starlings fly close together like an undulating, moving cloud, flowing all over the sky in an astonishing aerial dance. As many as 100,000 starlings have been counted in one murmuration in Britain.

INDEX

ABOUT THE AUTHOR AND ILLUSTRATOR

MICHAEL FEWER is an architect, environmentalist and writer. He has published work over the last twenty-five years on subjects varying from architectural history and travel to walking and nature guides. Previous books include *Day Tours from Dublin*, *By Swerve of Shore*, *Irish Waterside Walks*, *The Way-Marked Trails of Ireland*, *Irish Long Distance Walks*, all published by Gill Books, and *The Doorways of Ireland* (Frances Lincoln). He lives in Waterford.

MELISSA DORAN grew up in Leitrim and studied architecture and digital media technology before becoming an illustrator. Her favourite medium is digital collage that combines lots of hand-drawn elements. She also teaches digital illustration and organises sketching mornings in Dublin. To see more of her work visit www.goradiate.ie.

GILL BOOKS

Hume Avenue

Park West

Dublin 12

www.gillbooks.ie

Gill Books is an imprint of M.H. Gill & Co.

978 07171 6980 1

Designed by www.grahamthew.ie

Indexed by Eileen O'Neill

Printed by G.Canale & C. Spa, Italy

This book is typeset in 18 on 24pt Mrs Eaves Roman.

The paper used in this book comes from the wood pulp of managed forests. For every tree felled, at least one tree is planted, thereby renewing natural resources.

A CIP catalogue record for this book is available from the British Library.

5 4 3 2